Philip Crosby's

REFLECTIONS
ON QUALITY

∾

PHILIP CROSBY'S

REFLECTIONS ON QUALITY

295 Inspirations
from the World's Foremost Quality Guru

PHILIP B. CROSBY

McGRAW-HILL

New York   San Francisco   Washington, D.C.   Auckland   Bogotá
Caracas   Lisbon   London   Madrid   Mexico City   Milan   Montreal
New Delhi   San Juan   Singapore   Sydney   Tokyo   Toronto

Library of Congress Cataloging-in-Publication Data

Crosby, Philip B.
    Philip Crosby's reflections on quality : 295 inspirations from the
  world's foremost quality guru / Philip B. Crosby.
        p. cm.
    ISBN 0-07-014525-3
    1. Total quality management—Quotations, maxims, etc.   2.   Work
  groups—Quotations, maxims, etc.   I.   Title.
    HD62. 15.C763   1995
    658. 5'62—dc20   95–37400
                                                                    CIP

1 2 3 4 5 6 7 8 9 0   DOC/DOC   9 0 0 9 8 7 6 5

ISBN 0-07-014525-3

*The sponsoring editor for this book was Philip Ruppel, the editing supervisor
was Jane Palmieri, the designer was Marsha Cohen/Parallelogram Graphics,
and the production supervisor was Pamela Pelton.*
It was set in Bembo by Parallelogram Graphics.

Printed and bound by R. R. Donnelley & Sons Company.

INTRODUCTION

The first thing I learned about quality when I began working in an organization was that it was incredibly complicated. There were statistical concepts and applications that had to be understood in order to realize why it was necessary to plan on doing some portion of any work wrong. There was a whole formal system of determining what was "good enough" backed by government specifications to give it proper authority. The inevitability of practical error in the manufacturing areas was paralleled by a "that's close enough" mentality in administration and service. Quality was considered to be something that everyone knew when they saw it, yet everyone quarreled about it incessantly.

For the first year or so I assumed that I had quality all wrong. It had to be sensible, because so many people believed it. Books were written on quality control, seminars were held, departments inside companies were established, and there was statistical proof that nothing could get done properly the first time. Rework, by many names, was a normal part of doing business. MBA schools taught that quality was a trade-off: If you spent too much on it, you lost money; if you didn't spend enough, you lost money. They led the search for that "exact optimum."

My own thought was that we should learn to prevent problems by understanding the work, training people in its content, and aiming management at the practice of getting things done right the first time. I began to write and talk about this, only to be greeted with derision for speaking against what everyone believed to be absolutely true. I felt like Galileo explaining that the Sun did not revolve around the Earth. The proof is all about us, they would say; people can't do things right. I quickly learned that it wasn't possible to prove statistics wrong by using statistics as proof. So I began to see if I could express my thoughts in concise and easily digestible comments. The Book of Proverbs sat at the base of my mind. The brief, but true, comments compiled there have always been a source of pragmatic guidance to me in my private life.

My target for this conceptual revolution, really a reformation, was managers at all levels. They were the only ones who could change the way companies looked at quality. They had to realize that it was in the interest of their personal survival (the first of Crosby's Laws of Situation Management) to improve their operation by eliminating the waste entailed in doing things over and over and over. The management reaction was positive and supportive. Once they had a chance to think about it, my ideas made sense to them.

During my 27 years as a practicing quality management professional, I worked at all levels of organization from the assembly line to the corporate boardroom. Everyone I worked with let me try to change the place, and in most cases that happened with good results. However, in 1979, I realized that it was going to take a massive education effort to change the nation's and world's understanding of quality. The conventional wisdom was incorrect and damaging, to put it mildly. It believed that quality was "goodness," that it came from "detection," that the performance standard was "acceptable quality levels," and that it was measured by "statistical indexes." The result of all this was that companies spent at least 25 percent of their revenue doing things wrong—and over.

So I wrote *Quality Is Free*, which explained the subject to management in practical and achievable terms. When that book became a best-seller, I set up an educational organization (Philip Crosby Associates, Inc.) to teach executives and managers about the real world of quality. We did not advertise, and did not make sales calls. Everyone who came to the "Quality College" came on their own. Big companies like IBM, GM, Xerox, Milliken, PPG, ICI, Motorola, Chrysler, Johnson and Johnson, and dozens of others sent their top managers to class; hundreds of smaller companies like Tennant and Bama Pie came and changed their whole way of doing things. We taught 25,000 executives and managers each year in class, and hundreds of thousands of employees through the films and programs we provided for our clients.

What we taught them came primarily from the books I had written for McGraw-Hill. Recognizing that very few people read books from cover to cover, I created the "Guidelines for Browsers" in the back of each tome. Then we took what we came to call "Reflections" from the text and laid them out so that the reader could get an idea of the content of each chapter. When they wanted to know more about a subject, they could just turn to that page. Like the biblical Proverbs, each contains a thought behind the thought behind the thought. Book reviewers like the Browsers presentation, because it makes their job easier.

Quality—meaning, doing what you said you would do—is the only managerial aspect that pervades every area of every organization. In writing books on entrepreneurship *(Running Things)*, family management *(The Art of Getting Your Own Sweet Way)*, and organizational direction *(The Eternally Successful Organization)*, I used the concepts of quality management as a platform. All of what I learned came from dealing with the real world, and studying history at the same time. Each of these Reflections has a meaning to me. I hope many of them will be useful to you in your career and personal life.

Best regards,
Philip Crosby

PHILIP CROSBY'S

REFLECTIONS ON QUALITY

1. Genuine recognition of performance is something people really appreciate. They will continue to support the program whether or not they, as individuals, participate in the awards.

2. If quality isn't ingrained in the organization, it will never happen.

3. The executive's problem in understanding and utilizing the labor force is compounded by the fact that people are not interested in doing something just because they have been told to do it.

4. It is difficult to think of any profession you would want to be part of that wouldn't demand a great deal of integrity and compassion.

5. If a pearl of wisdom drops from your lips when no one is around, it will have no value.

6. Being part of a team is not a natural human function; it is learned.

7. Cooperation does not mean that you have to abandon any personal standards.

8. Being
understood is
hard work.

9. In the world of management, you *can* learn to be creative.

10. You can create solutions to complicated problems by being the only one to break those complicated problems down to their basic causes.

11. The most important aspect of creativity for management style is not getting into the habit of stating your lack of creative ability.

12. The most valuable manager
    is one who can first
    create, then implement.

13. People don't do for a living
    what they learned in college.

14. It is not possible to know
    what you need to learn.

15. Just about nothing is new, and almost everything used to be worse.

16. The leader who runs out of
    jobs for the led to do will soon
    be replaced by someone more
    interested in working hard.

17. If the leader is the only one
    who knows what game is being
    played, then obviously the leader
    is the only one who can win.

18. A good follower has to want the same results that the leader wants.

19. You never, *never* get away from being a follower.

20. Pretending all the time is a terrifying management style to adopt.

∾

21. *Listen.* You can convey no greater honor than to actually hear what someone has to say.

22. *Help.* Let someone lean on you without expecting to lean back.

23. *Create.* Original solutions are the result of hard work in uncovering unoriginal problems.

24. *Implement.* There comes a time when someone has to actually get the job done.

25. *Learn.* When you have an answer for everything, you know you have stopped learning.

26. *Lead.* Leaders begin to fail when they begin to believe their own material.

27. *Follow.* You will never reach a
stage when you aren't working
for someone, so learn to be
good at it.

28. *Pretend.* If you're going to be
an actor, be a good one. But
stay out of management.

29. "Zero Defects" is the battlecry
of defect prevention. It means
"do the job right the first time."

30. It is much less expensive to prevent errors than to rework, scrap, or service them.

31. Quality measurement is effective only when it is done in a manner that produces information that people can understand and use.

32. Most managers are so concerned with *today,* and with getting their own real and imagined problems settled, that they are incapable of planning corrective or positive actions more than a week or so ahead.

33. How you come across to
others should not be left to
chance.

34. Inside any organization, every
employee has a customer.

35. There is no substitute for the words "Zero Defects." They are absolutely clear.

36. If you don't know what the defect level is, how do you know when to get mad?

37. Many companies don't have the slightest idea what happens to their product or their service once it leaves their hands.

38. Quality is free.
It is not a gift,
but it is free.

39. Why spend time finding, fixing, and fighting, when you can prevent the problem in the first place?

40. Without active management, there can be no quality.

41. People perform to the standards of their leaders. If management thinks people don't care, then they won't care.

42. Getting people to understand the need for and learn to have confidence in an improvement program doesn't happen overnight.

43. The problem with using cash or financial awards to provide recognition is that they are not personal enough.

44. The cost of quality is the expense of doing things wrong.

45. Quality is ballet, not hockey.

46. The problem of quality
management is not what people
don't know about it. It's what
they think they *do* know.

47. The world's business, whether
a steel mill or a hotel, is
conducted by people.

48. Quality is hard to pin down,
because each person thinks
everyone else defines it the
same way he or she does.

49. To manage quality, you must define it as *conformance to requirements.*

50. To have a meaningful discussion on such a complicated subject as quality, some erroneous assumptions must be examined and altered.

51. *It is erroneous* to believe that quality means goodness, or luxury, or shininess, or weight.

52. *It is erroneous* to think that quality is intangible, and therefore unmeasurable. It is precisely measurable by the oldest and most respected of means— cold, hard cash.

53. *It is erroneous* to think that there is an "economics" of quality.

54. *It is erroneous* to think that all quality problems originate with the workers, particularly those in the manufacturing sector.

55. *It is erroneous* to assume that quality comes from the "quality department."

56. A person in the "manufacturing ghetto" can't contribute much to the prevention of problems, if all the planning and creation is done elsewhere.

57. Once integrity has been compromised, it can never return to its original pristine state.

58. The greater the distance between administrator and what's being administered, the less efficient the administration becomes.

59. If it's going to be practical and achievable, quality management must start at the top.

60. Prevention is harder to *sell* than it is to *do.*

61. Worker performance corresponds to the attitude of management.

62. Most of us live by a dual standard: one for our personal lives and one for our work.

63. What part
of "Zero
Defects"
do you not
understand?

64. People really don't work
for money.

65. Motivation's effect is only
short-term.

66. If you show no mercy for the
people who supply your needs,
why expect your customers to
treat you differently?

67. Quality has to be managed because nothing is simple anymore.

68. One of the most useless bits of management folklore: "If you have good in your heart, you will produce quality."

69. If you don't produce one dead dragon each week, your license may be revoked.

70. In every operation there's always one area that is more open to new ideas than the others.

71. The purpose of policies is to settle arguments.

72. One theory of human behavior is that people subconsciously retard their own intellectual growth.

73. The bigoted, the narrow-minded, the stubborn, and the perpetually optimistic have all stopped learning.

74. Customers deserve to receive exactly what we have promised to deliver.

75. It's always cheaper to do the job right the first time.

76. Many of the most frustrating and expensive problems we face today come from paperwork and similar "communications" devices.

77. Negative attitudes are more contagious than positive ones.

78. You could spend an entire career and still not experience all the different things that can go wrong in business.

79. Just because corporate has
    "gotten religion" doesn't
    mean anyone else has.

80. Anyone who considers quality
     improvement to be a motivation
     program will never attain
     Wisdom.

81. You can get rich by preventing defects.

82. An inspector is not a true inspector unless the inspection is independent and *last*.

∾

83. Once you put on a suit, no one tells you the truth anymore.

84. Good things happen only
when planned; bad things
happen on their own.

85. The safety of a product
isn't a legal problem, it's an
ethical one.

86. The basic cause of most product-safety problems is a lapse of integrity on the part of someone in management.

87. The first thing you notice when management moves into Enlightenment is the relaxation of tension.

88. If you don't expect errors, and are truly astonished when they occur, then errors just don't happen so much.

89. Imagine where your operation would be if you completely eliminated all costs related to failure.

90. We are a fragile and a vain group, we humans.

91. The reason for having an independent, objective quality department is to limit the making of decisions by those who have nothing to gain from them.

92. If something makes sense,
    is easy to understand, but still
    isn't done consistently,
    there must be a reason
    for not doing it.

93. You can make an excellent
    living actually doing the job
    of quality, as opposed to just
    auditing to find out why
    it *wasn't* done.

94. There is absolutely no reason
     for there to be errors or defects
     in any product or service.

95. Half of all rejections that occur
     are the fault of the customer.

96. People really like being
     measured, when the
     measurement is fair
     and open.

97. The only way senior management will let a quality team work on the future is if it is helping them to survive the present.

98. People will only tell you the troubles that others cause for them. They will not reveal the troubles they make for themselves.

99. Auditing catches only the
     undedicated, bored,
     or careless.

100. Millions of products are
     produced every day that don't
     wind up in court.

101. People just want their rights—
     until you try to trample them.
     Then they want revenge.

102. Companies that truly want to solve problems must create a culture of openness that is imbued with the basic concepts of integrity and objectivity.

103. Every human conflict produces winners, losers, and observers.

104. Over 85 percent of all problems can be resolved at the first level of supervision they encounter.

105. Once or twice in your career you will encounter something for which there is no solution. That's when you make a judgment call, accept the consequences, and get on with your life.

106. Specific
problems
require
specific
solutions.

107. "Assembly" is just making big things out of little things.

108. Quality improvement has no chance unless the individuals are ready to recognize that improvement is necessary.

109. Unless you know how you're doing as you go along, you'll never know when you're done, or if you've succeeded.

110. Anything that tastes good is bound to be bad for you.

111. Authority can be delegated, but only after it has been packaged.

112. Debt is not a friend.

113. One of the myths of management is that meetings are a gigantic waste of time. Only poorly managed meetings are not useful.

114. The price of nonconformance in manufacturing is more than your pretax profit.

115. We have traded Bob Cratchit's quill pen for a computer, but the process is still the same.

[handwritten annotation: Ebenezer Scrooge Employee]

116. We think
our customers
want us to
do things
differently.
We should
find out for
sure.

117. It is very easy to become complacent and slip away from communication.

118. Some "purposeful conflict" is necessary for success, but it can waste a lot of time if the purposefulness is not properly contained.

119. Setting forth a list of commandments will not convince anyone that management is serious about integrity. After all, integrity is as integrity does.

120. People like to think they shape events; in reality it's the other way around.

121. In the final analysis, their immediate supervisor is the person employees see as "the company."

122. Quality is free, but no one is ever going to know it if there isn't some sort of agreed-upon system of measurement.

123. Half the people in most manufacturing plants never touch the product.

124. We are all service people, except
     when we are blood donors.
     Then we are manufacturing
     plants.

125. The most difficult lesson for the
     quality crusader to learn is that
     real improvement just plain
     takes a while to accomplish.

126. Slowness to change usually means fear of the new.

127. You have to lead people gently
      toward what they already know
      is right.

128. There is a lot more to golf
      than having the equipment
      and the intent.

129. The people who have to put
     improvement programs of any
     kind into effect always feel that
     others are not for it.

130. No one likes to get out front
     with too much, unless they're
     absolutely certain it will be
     properly received.

131. Most things
don't work
like they're
supposed to.

132. The proof of understanding is
the ability to explain it.

133. No one is against preventing
defects; it's just that they don't
have the time right now.

134. If we take care of the customers
and the employees, everything
else takes care of itself.

135. Executives cannot be sold into being serious about quality; they have to make the decision themselves, and in their own time.

136. The primary measurement characteristic is how much money the company has made; the second is how long it has been around.

137. I had always assumed that everyone else knew a lot more about everything than I did.

138. We are all customers, and we are entitled to not be disappointed.

139. The criterion of being a
successful person is the ability
to get something done.

140. There are many cases where
the apparently invincible
organization becomes
vulnerable overnight.

141. Being convinced one knows
the whole story is the
surest way to fail.

142. Debt begins to rise; accounts
receivable age faster; bright
young people begin to drift
off to other companies; a huge
new headquarters building
gets built; customers get
neglected; housekeeping notably
deteriorates; and management
doesn't seem to notice.

143. Most organizations have
what appear to be suicidal
tendencies.

144. Organizations don't start out to be losers.

145. People can't last forever; the body eventually wears out. There is no way to prevent that from happening. But organizations have no need to die.

146. When the employees and suppliers of an organization do things correctly on a routine basis; when customers are pleased that their needs are anticipated and met; when growth is internally generated, profitable, and planned; when change is welcomed and implemented to advantage; and when everyone enjoys working there—the foundation of an eternally successful organization exists.

147. Executives are more likely
to be loyal to the company
and its products than to the
stockholders.

148. I cannot recall a single meeting
or conversation about doing
something for the stockholders
in all my years of working for
large companies.

149. Executives know they can move on to another company and not even be missed after a few months.

150. The eternally successful organization does not have to contend with uneven times, and is assured that survival is only an academic question.

151. Every business changes every day in some way.

152. Airlines are run by people who have never flown like normal folks.

153. The outstanding teams are invariably said to have accomplished their victories by "execution."

154. Very few of the great leaders ever get through their careers without failing, sometimes dramatically.

∾

155. We can spend our whole lives underachieving.

156. The great discoveries are
usually obvious.

157. Higher-ups have "mission
statements" which sound
good but are impossible to
measure.

158. Selecting the right person
for the right job is the largest
part of coaching.

&

159. Just being honest is not enough.
The essential ingredient is
executive integrity.

160. _Strategy_ is one of the most
overworked words in a
manager's vocabulary.

161. Management has to lead, but in some cases it has to drag people along.

162. Every time I hear the word *mission* I think of a big, complicated, political-sounding document that everyone nods about but no one takes seriously.

163. Grow where the business is; don't try to make a market where none exists.

164. Of course the customer comes first! Who doesn't know that?

165. Nobody's customers stay put.

166. Today's correct concepts are good only for today.

167. Top management should know
some customers personally.

168. No matter how it turns out,
I'm looking forward to
my brush with reality.

169. Look at change as being a friend instead of a pain in the neck.

170. Watch out for changes recommended eagerly by those who would profit from them.

171. Customers aren't as forgiving as our bodies are.

172. I would never send something to a customer that wasn't what I had promised to send.

173. Staying out of the quicksand is much better than getting a good deal on a towing contract.

174. It doesn't take long before the people running a place become convinced that it is their own personality and personal charm that cause everything to hum along so smoothly.

175. White-collar workers are about 50 percent effective.

176. "Tough" doesn't mean beating up on people. It means sticking to the necessary policies and actions no matter how enticing the reasons for easing up.

177. We all recognize that most of our problems are caused by management action, or lack of it.

178. Whenever management does something successful, or sorts out a complex situation, it too quickly forgets the whole thing and goes on to something else.

179. A rule to live by: I won't use anything I can't explain in five minutes.

180. When you decide to change a culture, you had better be sure you know exactly what you are going to change—and why.

181. The problem is usually in the process.

182. Quality is a choice that management can make.

183. Communication, Example, and Participation are the three legs of the platform for Pride.

184. Many people do not need much success to take something off the problem list.

185. It will take forever; but then, that's just how long we have.

186. Not everyone recognizes change when he or she sees it.

187. Quality has to be *caused*, not controlled.

188. Executives determine what is going to be run; managers do the running.

189. A manager's ability to get the impossible done just means that even more impossible tasks will soon be on the way.

190. It is not possible to fool people for very long. Once in a while they are wrong in judging those who are put over them, but they catch on very quickly.

191. If anything is certain, it is that change is certain. The world we are planning for today will not exist in this form tomorrow.

192. The business desert is layered with the bones of those who felt they understood completely, and stopped learning.

193. Making a wrong decision is understandable. Refusing to search continually for learning is not.

194. The best reputation a leader can possess is for displaying ethical conduct in all things.

195. Subordinates establish their personal determination level based on what they see in their leader. Wimpy behavior produces wimpy results.

196. Enthusiasm is the result of an energetic person working on something that he or she finds keenly interesting. This means that a leader should not sign on to any project that does not give him or her a tingle deep down.

∾

197. Nothing turns an organization ineffective so quickly as a continuing anxiety about what is going to please or displease the boss.

198. Running a corporation is too important a task to be left to the functional departments.

199. *Relationships* are where it all comes together or comes apart. Nothing else can be made to happen if relationships do not exist.

200. Wars very rarely break out
among those who have a
common understanding of
goals and purposes.

201. Quality is the result of a
carefully constructed cultural
environment. It has to be the
fabric of the organization, not
part of the fabric.

202. There is a great deal of evidence showing that change can be accomplished without spending much money or inflicting a great deal of pain. The solution comes down to a few simple understandings.

203. Quality means *conformance to requirements;* management's prime responsibility is to cause the right requirements to be created.

204. All action must be oriented
toward prevention.

205. Fortune 500 companies have
several more zeros at the
end of their numbers, but their
reporting and communicating
requirements are very much
the same as anybody else's.

206. Expenses are the key to profitability. There is no revenue stream that cannot be outspent.

207. It's hard to find a company that failed because its expenses were too low.

208. You don't have to be noisy to be effective.

209. This is a leader's job description:

a. Create the right environment—on purpose.

b. Reduce complex issues to something each person can understand and learn to handle.

c. Concentrate on the objectives of the operation.

d. Relate to people at all times.

210. Everything is personal. People don't want to separate home life and work life into two categories. You say it; they register it.

211. Don't go around asking permission to do what you already have been commissioned to do.

212. Eliminating what is not wanted or needed is profitable in itself.

213. ## Successful people breed success.

214. It's okay to be a fanatic about something, but only if that something can be explained quickly.

215. When it's a pleasure to come to work because the requirements for quality are taken seriously and management is helpful, then attitudes change permanently.

216. People should spend their time improving the quality process, rather than juggling it around to meet their feelings of the day.

217. Change should be a friend. It should happen by plan, not by accident.

218. It's hard to find an organization that both customers and employees regard with continuous affection and appreciation.

219. In a true zero-defects approach, there are no unimportant items.

220. The purpose of quality is not to accommodate the wrong things. It is to eliminate them, to prevent such situations.

221. It's hard to get people interested in improvement of any kind if they perceive it as a threat to their authority or lifestyle.

222. Never put development people in charge of production. They get too used to changing things whenever they feel it's necessary.

223. Quality is not a matter of having some superknowledge in some supergroup. It is a matter of managerial integrity. Either the requirements are taken seriously or they are not taken seriously.

224. There is too much emphasis on the ineffective aspects of quality improvement. Many people still think it's a technical problem, not a people problem.

225. I used to get a lot of free lunches by betting general managers that I could find someone violating a safety rule during our tour of their facility. It was always the general manager!

226. Requirements are answers to questions and the agreements that result from those answers.

227. We need to keep "requirements" in the proper perspective. It is we who are the masters, not the requirements. They serve to delineate an agreement between people, and should take whatever form is necessary. They must be respected and never altered, except by agreement among those who created them.

228. The product is the organization, and the organization is the product.

229. One of the most cynical statements imaginable is known as Murphy's Law.

230. I figure that every other person in service companies spends 100 percent of his or her time doing things over, chasing after data, or apologizing to someone.

231. All requirements come from the customer in one form or another, because with no customers there is no business.

232. The success of the quality improvement process does not depend on any "evangelical" powers possessed by the quality experts. It depends on education and implementation conducted in a serious and methodical way.

233. It always bugged me that someone I didn't choose got to make decisions about my career.

234. The quality improvement process is progressive. One doesn't just go from awful to wonderful in a single bound.

235. Each and every person in the
organization must understand
his or her personal role in
making quality happen.

236. Satisfy the customer, first, last, and always.

237. I start with two beliefs. First, those asking for help probably don't know much about the problem or they would be fixing it themselves. Second, there is someone there who *does* know, and no one will listen to that person.

238. Quality has always been the most negotiable of the "schedule, cost, quality" triumvirate.

239. "Zero Defects" means doing *what* we agreed to do *when* we agreed to do it. It means clear requirements, training, a positive attitude, and a plan.

240. The thought of error being inevitable is a self-fulfilling prophecy. If you think it has to be that way, it will be that way.

241. If we make our profit goals but don't pay our bills, then we have not met our profit goals. If we deliver on time but the product has defects, we have not delivered on time. If we meet our safety objectives but damage somebody, we have not met our safety objectives.

242. Improving quality requires an overall culture change, not just a new diet.

243. Listen to everyone; seek out
your customers and interrogate
them. Don't fall into ruts. Do
what you do thoughtfully, but
don't go to sleep. Someone
may be out in the garden
digging up your treasure.

244. No idea was ever accepted
right out of the chute.

245. Good ideas, based on solid concepts, have a great deal of difficulty being understood by those who make a living doing things the other way.

246. Trying to torpedo the competition should not be part of the business world. There is plenty of room for everyone.

247. Most real changes start somewhere in a company and spread around because they are worthy. They do not come out of the boardroom.

248. It only takes
one person,
one division,
one group
to change the
whole
company.

249. Relationships are what business is all about, and there are only two that matter: the one with customers, and the one with employees.

250. Someone has to be responsible for everything . . . or nothing will get done.

251. There are no failures in introducing quality improvement. No one ever gets worse, and no one ever doesn't improve.

252. Success is a matter of return on effort.

253. There is a Quality Revolution going on, and the quality people are missing it because they have their eyes on what goes out the back door, whether it's a product or a service.

254. It's hard to snuggle up to numbers.

255. The entire education process
can be summarized in
what I call "the six C's":
Comprehension, Commitment,
Competence, Communication,
Correction, and Continuance.

256. *Comprehension* is the
understanding of what is
necessary, and the abandonment
of the "conventional wisdom"
way of thinking.

257. *Commitment* is the expression of dedication, on the part of management first and of everyone else soon after.

258. *Competence* is the implementation of the improvement process in a methodical way.

259. *Communication* is the complete understanding and support of all people in the corporate society, including suppliers and customers.

260. *Correction* is the elimination of opportunities for error by identifying current problems and tracking them back to their basic cause.

261. *Continuance* is the unyielding remembrance of how things used to be and the glorious vision of how they are going to be.

262. Anything that is caused can be prevented.

263. The outgoing new products
     or services normally contain
     deviations from the published,
     announced, or agreed-upon
     requirements.

264. The companies that don't see much improvement, even though they appear to be determined, have common characteristics:

a. The effort is called a *program* rather than a *process*.

b. All effort is aimed at the lower levels of the organization.

c. The quality-control people are cynical.

d. Training material is created by the training function.

e. Management is impatient for results.

265. It has been known for generations that organizations that do what they promise to do, and take good care of their employees in the process, always come out on top.

266. There is absolutely nothing more demotivating or demeaning to a budding executive than to have to go to meetings where the assigned role is to be a faithful listener.

267. Dissatisfaction with the final product or service of an organization is called "trouble with quality."

268. The typical management decision that is the result of a "hunch" will cause quality problems.

269. Having a large book of policies and practices never saved any company from disaster.

270. The ecology of an organization is as delicate and vulnerable as that of a forest.

271. The performance review, no matter how well the format has been designed, is a one-way street.

272. Staff meetings are the same everywhere. The boss talks about what he wants to talk about for as long as he wants to talk about it, and then the meeting is over.

273. I never have felt that you could "motivate" anyone for more than a few days.

274. The first absolute:
*The definition of quality is
conformance to requirements.*

275. The second absolute:
*The system of quality is
prevention.*

276. The third absolute:
*Performance standard is
Zero Defects.*

277. The fourth absolute:
*The measurement of quality is
the price of nonconformance.*

278. The quality improvement
team requires a clear direction
and leadership. Otherwise,
people can get so involved with
strategy and the selection of
the team that they forget
what the team is *for.*

279. A great many people very rarely
have exciting days at work.

280. People don't work for companies; they work for people.

281. Shortcuts always cause problems.

282. We have to understand how to help the customers in ways they haven't thought about yet.

283. I have never met anyone who was against quality, or for hassles.

❧

284. The purpose of organizations is to help people have lives.

285. All employees want to be proud of everything they are connected with. If given the slightest opportunity, they will make their work something of which they can be proud.

286. I have never met an executive
who didn't claim to be
people-oriented.

287. The Navy and the Boy
Scouts know better.

288. Recognition must fit into the culture of the company.

289. Everyone expects the leader
to be all things.

290. There is no such thing as
a standard contract.

291. I don't know of any time of
life when a person can't
use counseling.

292. Very little can be accomplished without recourse to accountants, lawyers, and other business professionals. All of this creates a world of information understood in its entirety by no one.

293. Management is obsessed with fads.

294. Everyone picks a telephone system right the second time.

295. A leader has to be a reflection of the soul of the organization.